LOOK down on worlds before our world, at a time before our time.

See the great Ash Tree whose branches top the world of the gods and whose roots reach down into the land of the dead. Three worlds encircle the Ash Tree's mighty trunk, one above the other. The highest is Asgard, the land of the gods where Odin rules. From Asgard the Quaking Bridge, Bifrost, links the gods with Middle Earth where men like us live, and where dwarfs have their mines and tunnels. Below Middle Earth is the land of Hel, a place of dark and cold, and of the dead.

Look at Middle Earth. See leafy forests spreading over the land to the edge of the mountains that rise blue above the green of the plains. Around Middle Earth rolls the ocean, grey, misty.

Look now past Asgard, past Middle Earth, into the mists of mountain and ocean. See the Outer Lands, the cold kingdom of frost giants, Utgard.

It was a world of freshness and beauty. It should have been a place of happiness. It was spoilt and destroyed by rivalry between gods and giants. The forces of dark and cold from Utgard overwhelmed the sunshine and joy of Asgard and swept Middle Earth away with it.

The conflict that shook the worlds was fought out also in the heart of one who lived in Asgard: Loki.

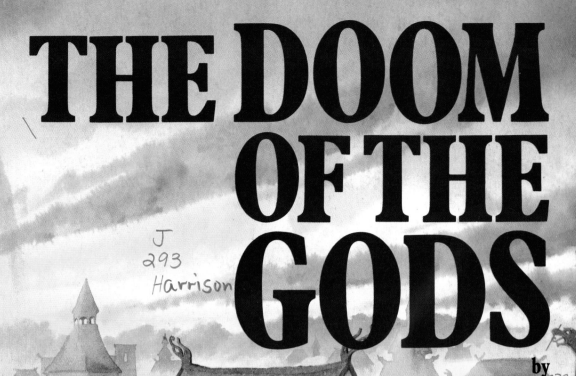

THE DOOM OF THE GODS

by
Michael Harrison

Illustrated by
Tudor Humphries

Oxford

Oxford University Press, Walton Street, Oxford OX2 6DP

Oxford London Glasgow
New York Toronto Melbourne Auckland
Kuala Lumpur Singapore Hong Kong Tokyo
Delhi Bombay Calcutta Madras Karachi
Nairobi Dar es Salaam Cape Town

and associated companies in
Beirut Berlin Ibadan Mexico City Nicosia

OXFORD is a trade mark of Oxford University Press

Text © Michael Harrison 1985
Illustrations © Tudor Humphries 1985
First published 1985
ISBN 0 19 274128 4

British Library Cataloguing in Publication Data
Harrison, Michael, 1985
The doom of the Gods.
1. Loki (Norse deity)—Juvenile literature
I. Title
293′.211 PZ8.1
ISBN 0-19-274128-4

Typeset by Tradespools Ltd., Frome, Somerset
Printed in Hong Kong

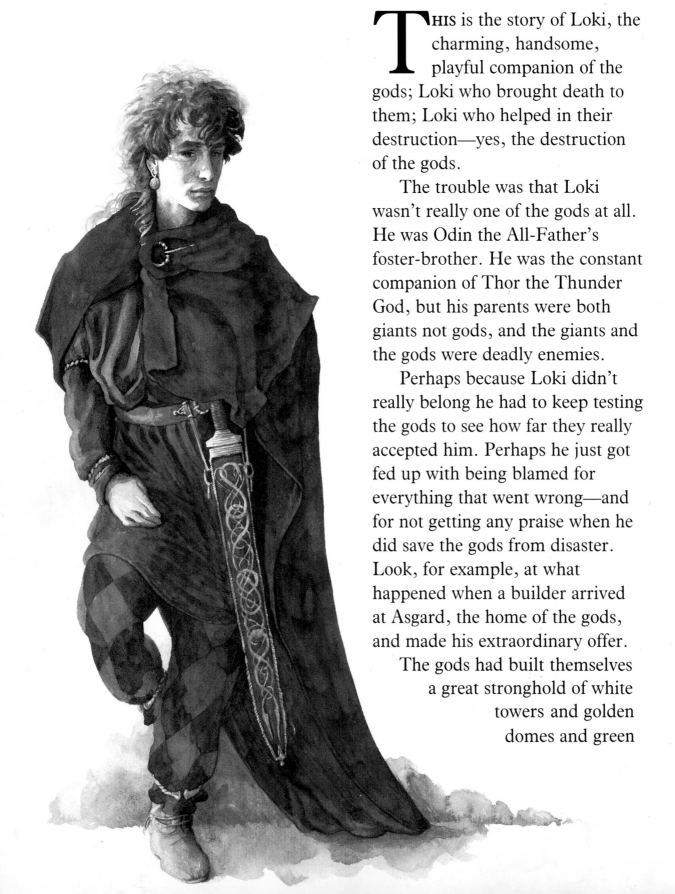

The Wall of Asgard

THIS is the story of Loki, the charming, handsome, playful companion of the gods; Loki who brought death to them; Loki who helped in their destruction—yes, the destruction of the gods.

The trouble was that Loki wasn't really one of the gods at all. He was Odin the All-Father's foster-brother. He was the constant companion of Thor the Thunder God, but his parents were both giants not gods, and the giants and the gods were deadly enemies.

Perhaps because Loki didn't really belong he had to keep testing the gods to see how far they really accepted him. Perhaps he just got fed up with being blamed for everything that went wrong—and for not getting any praise when he did save the gods from disaster. Look, for example, at what happened when a builder arrived at Asgard, the home of the gods, and made his extraordinary offer.

The gods had built themselves a great stronghold of white towers and golden domes and green gardens—and then they had given up. They had no wall around it to protect them from their enemies; no mighty gate at which the giants could hammer harmlessly. They preferred to spend their time feasting and drinking and telling stories. So when a stranger strode into their hall on the first day of winter, when they were beginning to think uneasily of the long darkness, and of ice and snow, and of the frost giants, they listened attentively. The stranger spoke knowledgeably about height and width, about the stone to use and the stone to avoid, about the dimensions and materials of the gate. He suggested subtly that they could relax behind such a wall and enjoy themselves without . . . not fear, of course, because the gods weren't afraid of the giants, were they? Without . . . he left it at that and moved on to describing the fittings of the gate.

The gods became impatient at this, as he had intended, and asked him his terms. 'Well,' he said slowly, 'let's say I've to finish it in eighteen months or I get no pay at all.' The gods all nodded happily at this for they could not see how one man could make such a wall in less than eighteen years. 'And if I succeed,' he continued, 'you will give me the goddess Freyja for my wife, and the sun and moon to keep us company.'

The gods were too shocked to speak. Freyja was the most beautiful of the goddesses and so it was natural that the builder would want her for his wife, that was quite understandable. But Freyja was the earth's mother; it was she who made the seeds grow in the spring and the fruit ripen in the autumn. With her gone, what would the earth be but desert? And the sun gone: no days, no warmth, no light. No moon to light the long northern nights.

Just before the gods broke out in a chorus of outrage one voice spoke softly, 'Why not say that the wall must be finished by the first day of summer, by dawn, by first light on the first day?' Loki's voice.

Now the gods were uncertain. They wanted the wall, without the work of building it themselves, but they could never pay that price. Surely, though, no one could build it in that time, in the months of one winter, with the ground frozen hard, the days so short. The builder said nothing, just let his mouth open and shut.

'Right,' said Odin, 'we accept your price, but you must finish by the first ray of sunlight on the first day of summer. If you fail you receive nothing.'

The builder muttered to himself. The gods were now crowding round him and he seemed to have no choice. 'I will need to use my horse to haul the stone,' he said.

There was more disagreement among the gods about this, but Loki spoke up again. 'Have you ever seen a horse that can build? If we want to get any wall built this winter he must use his horse. If he had to carry all the stone on his back we might get a couple of gateposts built by summer. They wouldn't give us much protection.'

'Very well,' said Odin, the father of the gods, 'You may use your horse. You have exactly six months to finish the wall. Finish on time and I swear in front of this whole company that you will be given what you have asked for. Fail, by just one stone, and you will receive nothing.'

'All right,' he said, 'I'd better start straight away.' The builder hurried from the hall. The gods laughed, returned to their drinking and felt they had been very clever. They slapped Loki on the back and drank toasts to their new wall. They told each other of the great deeds they had done and of their plans for next spring. And so, in familiar stories, in drink, they forgot the fearfulness of the bargain they had made. As it was

winter they stayed indoors out of the cold and time passed pleasantly enough. Outside the wind howled, and so did the wolves. The days were short. Snow, ice, cold, dark: who would want to go out?

If the gods had gone out their cheerfulness would have slid into dark pools of fear. The builder seemed to work with a giant's strength and his horse, a great grey stallion, dragged massive loads of stone, great blocks which the builder had cut out of the quarry as a farmer's wife cuts up slabs of butter for market. They were tireless. The great wall grew steadily. It was not just lumped together but carefully and neatly built. Standing outside, looking up at it through the twilight and the falling snow, you might have thought of a prison wall, grey and grim, and looked anxiously for the gate, but there was no gate yet.

At last, three days before the end of winter, some of the gods went out to see what progress the builder had made. They expected to see a few metres of rough wall. When they got outside they stopped, shocked. Stretching away on one side of a massive gate post was the wall in new white stone. To the left of the other gate post was a gap through which the pine trees could be seen. But the gap was small, a few metres only. The

9

raw edge of unfinished wall seemed to be stretching out cold fingers to the post. Even as they watched, the builder's great stallion galloped up dragging a load of stone. These mighty gods were like mesmerized rabbits watching a trap being constructed around them. As the stallion galloped off again the gods shook themselves and returned into the hall, white-faced.

They turned on Loki then, and blamed him. They said it was his idea, he had made them agree, against their wishes, to this mad bargain. They screamed at him, threatened him. Loki turned white, then turned on his heel and strode out. The gods jostled through the doorway after him. By the time they were outside Loki had disappeared. In his place an elegant mare was stepping daintily through the gap in the wall, lifting her feet high, twitching her ears, curling up her lip. She stood a moment, nervously. Then a great neighing came from down the hill and shouts from the builder. The mare ran lightly off towards the pine forest and the stallion was after her, the great sledge heaped with stone banging behind him. The gods could hear crashes coming from the forest as stone, trees, and sledge met, and smashed. The builder appeared in the gap. He was in a rage, a giant rage. He bellowed, cursed, stamped. As he raged he turned, before their eyes, into the giant he really was, the giant who had come in disguise to trick them. He stomped off to fetch stone by hand. There was so little left to build.

The gods hastily sent a messenger to fetch Thor the Thunderer who was travelling in the world of men, Middle Earth. When he arrived and had heard the story he laughed and said, 'Oaths gained by trickery and deceit are not binding,' and strode out clasping his hammer, the great Crusher. He came up to the builder as he was bending to place a stone in position. He raised his hammer and brought it down on the giant's head so that pieces of skull flew off and became embedded in the stones of the wall. With that load off their minds the gods placed the last few stones in position themselves and went back cheerfully to their feasting.

What of Loki? He returned to Asgard eventually, bringing with him a foal, for Loki himself had been the elegant mare. Loki was a Shape-Changer. He could change his shape at will to look like any person or creature, but he had to accept the consequences. Sometimes the consequences caused trouble but this time they were good. His foal was grey and had eight legs, four at the front and four at the back. The foal was swifter than any other horse and Loki presented him to Odin, the father of the gods. Odin called him Sleipnir, and thanked Loki. Most of the gods joked and offered Loki lumps of sugar. This was poor thanks, he thought, for saving Asgard.

Loki's Children

SLEIPNIR the eight-legged horse was just one, and not the strangest one either, of Loki's children. Though his goddess wife had two fine sons, sons that the gods would use later in their terrible revenge on him, Loki had three other offspring. The gods learnt of them through dreams and through the strange, half-waking dreams called prophecy. They learnt that Loki had had three children by a giantess, and they learnt that these three children would threaten them in the last days.

These dreams and prophecies were reported to Odin. Odin was the father of all the gods. Thor was his oldest son and the other gods were his sons and daughters too. Odin was god of war and was accompanied by wolves and ravens. He determined to defeat the prophecies by decisive action. He sent a group of younger, bolder gods to seize Loki's offspring and to bring them back to Asgard. The gods journeyed to the land of the giants, Utgard, that lies over the mountains to the east of Middle Earth. They broke into the giantess's house at night, tied her to her bed while she lay snoring, stuffed her great stocking into her mouth to keep her quiet, grabbed three squirming creatures, thrust them into sacks, and left as quietly and as quickly as they could.

They returned to Asgard and laid the sacks at Odin's feet. The father of the gods stepped down from his high seat and pointed to the left-hand sack. The nearest god bent down, untied the strings and tipped out a writhing snake. It lifted up its head and hissed at Odin as its forked tongue flickered in and out. Its scales rasped as it coiled and uncoiled. Odin picked it up by its tail and, with a quick flick of his wrist, sent the snake spinning down from Asgard, down, down to Middle Earth. Here it splashed into the great ocean that surrounds the land. Here, nosing around on the bottom of the ocean, the snake grew larger and larger until it felt its own tail growing down its mouth. Now it circled the whole of Middle Earth and lay around the land like an enormous belt. This was Jormungand, a great serpent waiting under the surface of the sea.

When the second sack was tipped up there tumbled out a baby girl, quite naked, quite repulsive. She was half pink and half black but, when you looked closely, the top half was blue and goose-pimpled and the

bottom half was the colour and texture of a rotting corpse. Her face scowled up at the surrounding, shuddering gods. Odin picked her up by one wrist, swung his arm, and threw her arching over the edge of Asgard, past Middle Earth, into that land of the dead that is for those who die painfully of disease or old age. A cold foggy land with a cold gloomy hall, full of cold and sullen spirits. Here the baby grew, and grew more surly

and more ugly. She ruled over that land, and land and ruler grew more and more alike so that the land is named Hel after her. To Hel went all the dead except for the greatest heroes who fell in battle. Remorse ate into their minds as worms ate into their bodies. Hel sat on her throne of bones waiting for her father's summons.

And the third sack? A puppy stalked out. Odin bent, hesitated, straightened. It was a wolf puppy. Odin was god of battles and the raven and the wolf, scavengers among dead bodies, were his special creatures, sacred to him.

'He'll stay here where we can keep an eye on him,' he said. 'His name is Fenrir.'

Fenrir was never a playful puppy, in fact he was not a puppy for long. He grew fast and his temper, which was bad to begin with, grew worse too. None of the gods wished to go near him. Only Odin fed him, threw him the bones and hunks of bloody flesh. Even Odin paled when he saw those teeth tear into the meat and crack open the bones. Fenrir panted as he ate, saliva ran over his lower jaw and his red-rimmed eyes darted round. So months passed and so Fenrir grew, and so the gods' fear grew too.

At last even Odin agreed that Fenrir must be bound, chained up so that he would be under control. They sent to the dwarfs for a strong iron chain. It was no use trying force against such a strong beast, so they tempted his pride.

'It is time,' they said to him, 'for you to make a great name in all the worlds, here in Asgard, in Utgard the land of the giants, in Middle Earth among men, and in Valhalla and Hel, the lands of the dead. Prove your great strength by breaking this iron chain.'

Fenrir sniffed at the chain suspiciously but his conceit was greater than his caution and he agreed. Odin himself tied the chain round him, tested his knots, stood back, and said, 'Try now, Fenrir!' The great wolf arched his back, and then, before the gods saw quite what had happened, Fenrir stood shaking himself while the chain lay on the ground. After a moment's shocked silence they hastily praised his great strength and examined the chain. The links were stretched thinner, some were stretched to threads, broken threads.

Odin immediately ordered the dwarfs to make a stronger chain.

When this arrived they took it in again to Fenrir. 'Everyone is talking of your great strength,' they said, 'and this chain, which you see is stronger than the first, has been sent by some of your admirers to enable you to establish beyond question your mighty power.' Fenrir saw that the new chain was very much thicker and heavier than the first, but he knew too that he had been growing steadily—and his conceit had not weakened either.

So again Odin fastened the chain round the grey beast, even more carefully this time. Then he stood up, a little anxiously. Fenrir arched his back, dug his legs in, put his head down and, slowly this time, the links lengthened, thinned, and parted. Fenrir panted slightly, curled his lips at the gods, and demanded food.

Odin in his fear sent a messenger to the dwarfs saying, 'Make me a rope that will bind Fenrir. If you fail then I will send Fenrir down to earth to feed off you. He will lick you up with his great red tongue and swallow you whole.'

The dwarfs knew that no metal would hold the wolf now so they made a light, silky cord by plaiting together six threads: the noise a cat makes when it moves, the beard of a young woman, the roots of a mountain, the sinews of a bear, the breath of a fish, and the saliva of a bird. The dwarfs brought the finished cord to Odin.

Odin and all the other gods brought Fenrir this time to a small rock in the middle of a lake. 'You have now twice shown your great power,' Odin said to him. 'The third time will establish your name for ever as the strongest of all creatures.'

When Fenrir saw the cord he was immediately suspicious and backed away. 'Something as light as that must be magic,' he snarled. 'I'll not match my strength against magic.'

Odin assured him that there was no magic at all but Fenrir was not convinced. 'If one of you will place his right hand in my mouth, then you can bind me,' he said. Odin looked around the circle of the gods, and they all looked at him. No one spoke. At last Tiu, the gods' most mighty fighter, strode forward and held out his right hand. Now the gods' hopes were divided. If they succeeded in binding Fenrir they would be safe from him, but be weaker by the loss of Tiu's hand. Odin did not hesitate. He knotted the cord round each of Fenrir's legs and fastened it to the rock. Fenrir strained so that his great muscles stood out, his eyes bulged in his head. All the time Tiu stood, his arm out, his hand in Fenrir's frothing mouth, with its curled lips and white teeth.

Fenrir's stuggles succeeded only in tightening the knots and he gradually weakened. Despair crept into his heart. Suddenly he stopped struggling and snapped his mouth tight shut. He swallowed once and there stood Tiu, his right arm streaming blood, his hand gone. The gods bound up the stump and led him away while Fenrir stood, howling to the darkening sky. And there he stayed, until the last days. Tiu was never again the greatest warrior, or not until the last great battle, but his name was held in great honour and Tuesday was named after him.

Where was Loki? Loki was on a journey with Thor.

Utgard

THOR and Loki were away from Asgard, travelling through the worlds to see what they could see. Thor took Loki with him because Loki's humour and cheerfulness made the journey pass more pleasantly. They had reached Middle Earth, the land of men, and came to a lonely region of moorland. The ground was boggy and they had to get out of Thor's chariot and walk beside the two goats who pulled it. The sun made pink and green patterns in the sky and the air grew chilly. Each time they reached the crest of one hill they saw another rising above them. At last, as darkness grew more solid, Loki saw the red glow of a fire flickering in the distance. As they drew nearer they saw it was by a hut made out of squares of turf cut from a field in which a thin crop of barley now grew. At the sound of their approach a boy of about eleven and a girl a little younger looked out from the doorway. They gaped, went back in, came out behind their mother. The father came round the corner of the building, and he too stopped. They had obviously recognized the travellers: Thor by his mighty hammer and Loki by his red hair. They stood stammering, wondering what trouble this meant.

'Peace be with you,' said Thor, untying his goats. 'We will shelter here tonight. Prepare a meal for us.'

'But . . .' began the woman.

Loki's quick eyes had absorbed the poverty of the hut and had seen the watery barley-gruel cooking over the fire. He saw, too, the careworn faces of the parents and how the bones stood out in the children's faces. His impetuous nature led him sometimes into mischief, sometimes into kindness. He now led Thor a little distance away and told him that these people did not have enough, even for themselves. Thor laughed. He took out his knife and cut the throats of his two goats. He skinned them carefully and gave the meat to the woman to cook. Soon the smell of the roasting meat filled the hut and soaked into the walls. In the months ahead if the family leaned on the walls of their hut they were filled with the memory of that meal.

When the meat was cooked Thor spread the goatskins on the floor, hairside down. 'As you chew each bone clean,' he said, 'throw it on to a

skin. Do not crack or break the bones or Crusher will crack and break you.' He waved his great hammer at them as he spoke.

At first the family were too frightened to eat, but gradually they started on the meat. Thor and Loki had great appetites and devoured the greatest part and the parents had most of what was left. The boy, Thialfi, seeing this wonderful food disappearing into the mouths of others, quietly picked up a bone. He cracked it and sucked out the golden marrow. He threw the bone on the growing pile and spooned up his gruel. At least there was enough of that tonight. When the gods had finished they wrapped themselves in their cloaks and lay down next to the fire. The family lay, uncomfortably full of food, in the corners of the hut.

In the morning, mist had spread everywhere and its clammy breath covered the moors. The gods woke, stiff and cold. They munched silently at the barley biscuits the wife offered. Then Thor took his hammer and held it up like a cross over the pile of bones. He spoke a few words of the old language and the bones moved. They fastened themselves together, swelled with flesh, and were wrapped with skin.

The two goats got to their feet as Thor smiled. But his smile was soon lost in a yell of rage as one of the goats limped out through the door. He swung his hammer and was about to smash them all when the boy Thialfi fell forward, clung to Thor's knees, and sobbed out an apology. Thor lowered his hammer and thought. He did not have Loki's quick mind, in fact he thought so slowly he could appear stupid. His answer to most problems was a crushing blow but this time two things made him hesitate. He was always fair and just by his own rough standards and he had an enormous appetite for food and drink. He remembered the supper he had eaten the night before and he remembered the poverty of the family. Then he remembered his own stern warning. He laughed.

'You two children will come with me and be my servants. I will leave

these four-footed goats here for a while. They may be more use to your parents than their two-footed ones.' Fair, perhaps, but the god seemed unaware of the strength of human love, of family love. The children's parents watched sadly as the four walked off through the mist.

The countryside turned from moorland into thick pine forest. The trees were so close together that, when the sun dissolved the mist, only a few rays reached down to the ground. The lower branches of the trees were all brown and the floor of the forest was thick with pine needles. It was silent too; few animals or birds lived there. It grew dark early and the children began to look about nervously for wolves and shadows. Their mother had often told them the story of the great wolf that chases the sun across the sky every day, snapping at his heels. As the sun grows tired the wolf's teeth catch in his flesh and his red blood spreads across the heavens.

As the last light of the sunset faded from the sky they came to the mouth of a cave, a deeper blackness in the gloom of the forest. Thankfully they went into it, put down their luggage and settled themselves for sleep.

They had not been asleep long when the ground shook and the air was filled with peals of thunder. They ran deeper into the cave and found a smaller opening off the main cave. They went into this and huddled at the end of it until morning as the shaking and thundering went on all night.

At first light they crept out of the cave. Ahead of them, through the trees, they saw the sole of an enormous boot pointing up to the sky. They edged towards it and looked round it. Stretching away from them and filling the clearing was a giant, flat on his back, snoring. He suddenly stopped snoring, opened his eyes, and looked straight at them.

'Ha!' he said, and spat out a puddle. 'So, Thor, you're on your way to Utgard, are you? What do you want in the land of the Giants?' As he spoke he got to his feet and a strange thing happened. He seemed to be smaller standing up, a giant indeed, but not the mountain range he appeared before. 'Ah, there it is!' he said, bending down and picking up a

large mitten from the ground, from where they had spent the night. There was no cave there at all now, and that was strange too.

'Who are you?' demanded Thor. 'How do you know where I am going better than I do?'

'Me, they call me Skrymir, or the Big Fellow,' the giant said. 'Breakfast, I think.' He sat down and untied a sack he had with him. He took out joints of cold roast meat and ate them noisily. Thialfi opened the bag he had been carrying and passed some of his mother's barley biscuits to Thor, Loki, and his sister. Skrymir drank from a skin bottle of red

wine while the others found a spring and drank the icy water.

Skrymir was tying his sack up when he paused and said, 'Let me help you on your journey. We'll share supplies if you like. Shove yours in here and I'll carry them.' They didn't hesitate a moment, thinking of the meat and wine. Skrymir tied the cords round his sack and flung it over his shoulder.

'Come on, little ones,' he said and strode off through the forest. Thor the mighty Thunderer was reduced to stumbling at Skrymir's heels like a reluctant child. They walked all day without a rest and by nightfall had come to the edge of the forest. They sat down under one of the great oaks that flourished beyond the pine trees. Here Skrymir tossed his food sack at Thor.

'Eat, little ones,' he said, and immediately lay down and started snoring.

Thor tried to undo the knots of the sack but the cord seemed to be made of iron. The more Thor struggled to loosen them the tighter they seemed to get. Thor was very tired and very hungry. He was becoming very angry. Loki and the two children watched his hands eagerly but did not dare to speak. Hunger ran in their bellies but fear of Thor sat on their tongues. At last Thor flung the bag down and pulled out his mighty hammer, Crusher, from his belt. Gripping it in both hands till his knuckles showed white he swung it over his head and brought it down with all his strength on the giant's forehead. Crusher bounced on the bone. Skyrmir stirred. He rubbed his forehead, and sat up.

'This tree is shedding its leaves on me and keeping me awake,' he said. 'We'll move under another one. Have you little ones eaten yet? Why aren't you asleep?'

Thor and Loki hastily assured him, even as their bellies rumbled, that they had eaten and were ready to settle down. When Skrymir had settled himself down under another oak and started snoring again, Thor quietly studied his face. He was looking for the best place for his next blow. He raised his hammer, and crashed it down. This time the head of Crusher sank into Skrymir's head, leaving a hole Thor could put his fist into.

But again Skrymir sat up and cursed. 'This oak is dropping acorns on me now. What are you doing, Thor, standing there? Are you trying to catch them before they land on me?'

Thor hastily explained that he was just going to sleep but he was looking for the best spot. In his heart he resolved that his next blow would be his last. He sat and watched until the first rays of dawn lighted on Skrymir's head. He swung his hammer again and brought it down with such strength that it went in up to the handle. He pulled Crusher out and was getting ready for another such blow when Skrymir yawned deeply, sat up and said, 'There must be pigeons roosting in this tree. I felt a bird-dropping splash on my face. Now, my little ones, Utgard and the Hall of the giants lies just ahead of you. Keep straight on, and mind your manners. If you think I'm big, wait till you see the real giants. They don't take kindly to cocky little men.' He threw down his food sack and strode off. His giant strides seemed to take him out of sight immediately.

Thor kicked the food sack angrily. Roskva, the girl, was so hungry that she could not stop herself trying to undo the knots. Her quick fingers had it untied in a moment. Inside were just their barley cakes and water-bottles—no meat, no wine—but this was feast enough for their starving stomachs. Strengthened by this breakfast Thor determined to follow the giant to Utgard.

They had not been walking for long when, there, straight in front of them, reared the massive icy walls of Utgard, the stronghold of the giants. Thialfi wondered why he hadn't noticed the walls before in that flat, bleak plain, scoured by the wind. He guessed it was because he had been lost in thoughts of home. All the others seemed surprised, but no one said anything. They came to great iron gates which swung open by themselves as they approached. After a moment's hesitation they walked through, Thor gripping his hammer; all that is except Roskva. She was sure Utgard had not been there five minutes before and would not pass the gate. She sat outside to wait.

Thor, Loki, and Thialfi walked on through silent streets between huge buildings that shone with an icy-blue glow. They walked until they saw light and heard noise coming from an open doorway. As Thor stepped through there was silence. A hall full of giants turned instantly and stared at the newcomers. Thor strode forward towards the high table and spoke to the giant who sat there.

'Are you the Lord of Utgard?'

The giant laughed. 'You might prefer to call me the Loki of Utgard perhaps. But yes, I am the Lord of Utgard. You must be the midget Thor I have heard about.'

Thor kept his temper and smiled tightly. 'I am Thor, the Thunderer, and I bring you greetings from Odin, the Father of All the Gods and Lord of Asgard. I come in friendship.'

'Welcome, then, Thor. We have a custom in Utgard. All our guests entertain us by showing off their strength or any particular skill they have. What can you and your friends offer us?'

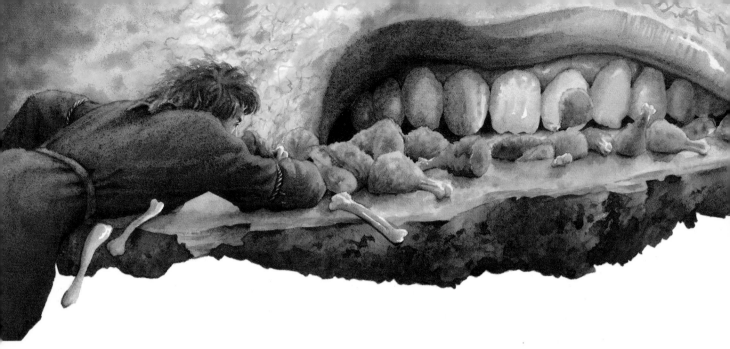

At once Loki pushed forward, his red hair catching the light from the flickering flames of the torches.

'I am a mighty eater,' he boasted. 'None of you, big as you are, can eat as fast as I can.'

The Lord of Utgard smiled. 'This will be a contest worth watching,' he said. He called to one of the giants. 'Here, Logi, you are our best eater. Match yourself against our guest, Loki.'

Two of the giants prepared a long dish of meat. The dish was made of wood, a slice from down the trunk of an oak. On it they placed joints of roasted goat. Loki and Logi stood, one at each end, waiting for the signal to start. No sooner did they start than they seemed to meet in the middle. Loki stood up triumphantly, and at least his stomach was full. He was soon humiliated though. The giants broke out laughing and pointed to the table. There lay Loki's half of the plate, with the bones picked clean. On the other half of the table was . . . nothing. Logi had eaten meat, bones, plate.

'Well tried, little Loki,' said the Lord of Utgard. 'Now, what can this midget midget do?' He pointed at Thialfi.

'I am a fast runner, Lord,' said Thialfi.

The Lord of Utgard called over a slender, frail-looking young giant. 'This is my—my son Hugi. He will race you.'

The giants and their guests went out of the Hall. There, outside the door, was an open space that Thor and Loki had not noticed before. Two running tracks were marked out. As the first race started they could see that Thialfi had spoken truly. He was very fast. But, fast though he was,

Hugi had returned to the starting line before Thialfi had reached the half-way turn. The same thing happened a second and third time, until Thialfi gave up and admitted he was defeated. The giants returned to the Hall in triumph. Thor was scowling now for he knew his turn had come. The great gods do not enjoy humiliation.

'Now, Thor,' said the Lord of Utgard, 'what party-piece do you have?'

Thor stood straight and replied, 'I can drink.'

The Lord of Utgard clicked his fingers and a giant came forward with a drinking-horn that looked big even when carried in a giant's hands. Thor grasped it and saw that the wine inside lapped at the edge so that it was almost overflowing. He put the horn to his lips and drank deeply. As he lowered the horn after his first great gulps he saw that the level had

dropped, just. He put the horn to his lips again and swallowed as deeply as he could. This time he lowered it confidently.

'You have done well if you have emptied it in two, Thor. My giants take three goes before they succeed.'

Thor looked down and was dismayed to find how full the horn was. He raised the horn for his third attempt and drank until he felt his eyes would burst out. He lowered the drinking horn. All he had managed to do was to lower the level by the thickness of his thumb. The giant who had brought the horn to him took it away again and showed it to the Lord of Utgard.

'Well done indeed,' he said to Thor. 'Now that you are refreshed, would you please pick up my cat and pass it to me.'

Thor smiled gamely under these insults. A large grey cat paced into the Hall. It came and stood in front of Thor and slowly turned its head towards him. Its green eyes stared with placid contempt. Thor bent, seized the cat round the middle and tried to straighten up. He strained until, finally, he arched the cat's back and one paw was off the ground. He could do no more. His own great sigh as he dropped the weight of the cat covered the gasps of surprise that broke from the watching giants—except from Loki. Loki's quick ears caught the sound, his quick eyes caught the expressions of dismay on the giants' faces, and his quick mind guessed at the truth.

Thor turned, red with exertion and anger, as the Lord of Utgard spoke again. 'Well, my little one, if my cat is too heavy for you would you like to try wrestling with my grandmother?'

As he spoke an old woman hobbled into the room. She was so old that her face seemed to have turned into a skull. Thor paused, thinking of the drinking-horn and the cat, but the mocking cries of the giants provoked him into the attack. He rushed at the old woman, intending to raise her into the air above his head and smash her down at the feet of the Lord of Utgard, splintering her bones.

He did not manage to lift her off the floor by the least bit. He grunted and struggled. He exerted all his strength but the old woman inexorably, slowly, forced him down, down, until one of his knees touched the floor.

The Lord of Utgard sprang to his feet and ran towards them. He pulled the old woman from Thor and helped him up.

'You have all entertained us enough,' he said. 'Come and join our feast.'

They feasted far into the night and then lay down in the great Hall and slept.

In the morning the Lord of Utgard led Thor, Loki and Thialfi outside the gate. There they found Roskva waiting for them.

'You were not long,' she said to her brother.

Thor thanked the Lord of Utgard for his hospitality but he just laughed at Thor's dejected expression.

'Do not feel humiliated, Thor,' he said. 'We knew you were secretly coming to destroy us. If we had known how strong you really are we would have kept you well away from Utgard. You have been tricked and deceived from first to last. I was Skrymir, whom you met in the forest. I made my glove into a cave for you. I protected myself by magical illusions when you struck me with your mighty hammer. On your way back to

Asgard you will see three craters in the ground. You made them when you thought you were hitting me.'

He turned to Loki. 'Your contest was hardly fair. Logi was fire, and who can eat faster than fire? And you, Thialfi, you raced against my thoughts. And thought can circle the world in no time at all.'

Loki smiled to himself. His guess had been correct.

'What tricks did you use against me?' growled Thor, realizing now why the Lord of Utgard jokingly called himself the Loki, the Trickster, of Utgard.

'The drinking-horn I gave you had no bottom,' the giant replied. 'It went down into the sea. When you drank so deeply you started the ebb of the sea that men will call tides. They will say Thor Is Drinking to Ebb the Sea. The cat you nearly lifted was the great serpent Jormungand that Odin himself threw down and which now circles the world. You stretched it up so that it nearly touched the sky and brought the heavens smashing about us.'

'And the old woman, your grandmother?'

'She is old age. No one can win against her.'

The Lord of Utgard turned, and as he did so he said, 'You will never come here again. You and I, Thor, will meet at the Last Days, at the Twilight of the Gods.'

Then, he was gone, and Utgard was gone, and they were left on a vast and windy plain. They set off silently to return Roskva and Thialfi and to collect the goats, Thor swinging Crusher thoughtfully. Perhaps he was thinking of how he would like to hammer the giants and their Utgard flat into the plain. Perhaps he was remembering how Loki's tricks and jokes had brought Crusher to him years before.

Sif's Hair

No one knew how Loki had come to live with the gods. There were rumours, stories that the gods told in lowered voices when Odin wasn't around. If Loki wasn't a god he wasn't exactly a giant either. With his flaming red hair he seemed to belong nowhere. All that was certain was that he had been brought up as Odin's foster-brother. In fact, he was Thor's constant companion, until a coolness grew up between them. It had happened years before and it had been Sif who had caused it.

Thor's wife, Sif, had long fair hair. She used her hair to tease her husband's friend Loki. She would turn suddenly when Loki was near so that her hair swung out into his face. Then she would laugh sweetly and say how much better it suited his handsome face than the red of his own hair. It was, perhaps, amusing once or twice but even good jokes pall and this joke had unkindness behind it.

Loki smiled and smiled for his friend Thor's sake until, one evening, his mischievous spirit got him into his first serious trouble with the gods.

Loki had gone into Thor's bedchamber to fetch something for his friend. In the flickering firelight he saw Sif asleep on her bed. Her hair, her golden hair, was spread out over the pillow, catching the red light from the fire. Without thinking, Loki picked up Sif's scissors and cut off all her hair so that her head looked like a cornfield after the harvest of the golden stalks. He sat down by the fire and began to weave the hair, trying

34

to make a wig for himself. He thought
he would show Sif how he would look
with fair hair.

It was here that Thor found him
when he came to discover why he was
taking so long. There was Loki, sitting
cross-legged in the firelight,
concentrating on something in his lap.
Thor smiled. He was used to his
friend's impetuous, forgetful nature. He
turned from his friend to his wife. It
took some moments for him to
understand what he saw, before he
connected his wife's close-cropped hair,
her suddenly altered face, with the
yellow strands in Loki's lap.

Thor roared, Sif woke and
screamed, Loki cried out in terror.
Thor seized Loki, swung him high in
the air—and hesitated. Perhap he
could not decide whether to smash him
on the stone slabs of the floor until his

35

red blood painted them, or whether to fling him into the red flames of the fire so that they would leap up yellow with the extra fuel. Perhaps he paused out of old friendship. Perhaps he just wanted to savour his revenge and enjoy the moments before he killed Loki.

Whatever the reason, a pause was all Loki's quick tongue needed.

'My dear Thor,' he said, quickly, but not too quickly, 'I am—if you put me down—on my way to the dwarfs who live under the earth. Sif deserves hair made of real gold, not this pale imitation. They will make

hair for her. I need this for them to match the colour. And of course,' he added, as Thor still hesitated, 'her gold hair will grow like real hair, and will need cutting . . .'

Few are free of the greed for gold. Thor tossed Loki down, but not too roughly.

'See you return quickly,' he growled.

'Of course,' said Loki, picking himself up and edging from the room.

'And remember,' shouted Thor after him, 'it is the custom for travellers to bring presents back to those they leave at home.'

So Loki set out from the high bright halls of Asgard, down over Bifrost, the rainbow bridge, to Middle Earth, the cloudy land of men. And from Middle Earth he went down into the earth itself.

The dwarfs lived in the earth like maggots in flesh, for they had once been real maggots. They had first eaten their way out of the Ice Ogre, Ymir. At the beginning of the worlds Ymir had been formed from the meeting of Ice and Fire and from his broken body the worlds were made: seas from his blood, mountains from his bones, trees from his hair, and the sky from his skull. The gods saw the maggots crawling, bloated and fat, from Ymir's flesh and gave them the shapes and minds of men. They set them in mines and underground tunnels. Here they dug, and refined, and worked the metals the gods and men needed.

Loki came to the entrance of the dwarfs' world. It was a cave set into the side of a hill. Sunlight reached a little way into it but then it became darker and lower. It was difficult for one from the outside to enter, difficult to face the thick blackness of that unknown world, with its invisible dangers.

Loki walked bent double, and crawled, and banged his head and scraped his elbows, through miles of tunnels. He was searching for Ivaldi, the most cunning of all the dwarfs. Dwarfs he met were not helpful. One, Brokk, sneered and said that Ivaldi was a mere black-smith.

'My brother Eitri is the only real craftsman here. You would do well to deal with him.'

Loki ignored him and went on.

Finding Ivaldi was hard, persuading Ivaldi to make what Loki so desperately needed was easy, for Loki. His quick mind saw that what the dwarfs in their dark underground tunnels craved most was praise and renown in the world of light. Loki's quick tongue painted such a picture of the glory he could win for the dwarfs that Ivaldi and his brothers fired up and promised Loki not only the hair he needed but two other wonders as well.

The dwarfs worked at their forge, pumping the bellows, hammering and muttering. Loki stood in the flickering shadows at the side of the cave and watched the finest threads of gold shining purely in the heat of the red fire. The dwarfs' shadows loomed huge and the concussion of their hammers bounced around the walls. At last they were finished and Ivaldi placed in Loki's eager hands strands of the finest gold, finer than human hair.

'Place this on Sif's head,' Ivaldi said, 'and it will grow and look like her own hair. Now, let me see what else I can find for you.'

Loki stood over Ivaldi, exclaiming at the workmanship and describing the scene in Asgard when he would place the gold on the astonished Sif's head. How Odin, the Father of all the gods, would rise slowly from his throne. How Thor would thank Loki for cutting off Sif's hair. How the other goddesses would finger their own hair. How all would say, 'The dwarfs did that?' How the giants would think of excuses to visit Asgard to see Sif's hair. How men would tell of the wonder around their fires on Middle Earth.

Ivaldi, smiling broadly, presented Loki with two objects. The first was a plain spear.

'What great work of the dwarfs is this?' asked Loki.

'This,' Ivaldi said, 'is Piercer. It will never miss its target when you throw it.'

Loki was silent when Ivaldi produced a flat package, waiting for his explanation.

'This is Skidbladnir,' said Ivaldi. 'As you can see, it folds up like a handkerchief and can fit into your pocket. Unfold it, and it is a boat that you and all the gods can use to carry you across the seas. Not only that, but its sails will catch a wind as soon as they are hoisted. Take these to Asgard and show the gods what we dwarfs can do.'

Loki's thanks were genuine, for he was sure of buying his way back into Asgard now. He left the dwarfs glowing with pride and exertion and

set off for Asgard. The journey back through the tunnels and mines was no easier than the journey in but Loki's heart was lighter. As he passed the cavern of the surly Brokk again he stopped and said,

'Look what the blacksmith has knocked up for me.'

Brokk spat. 'Clumsy stuff. Stuff like that will give the dwarfs a bad name. Leave it here, please.'

'That is your jealousy,' said Loki. 'Your brother could not do work like this, my head on it.'

'And who will judge between us?' asked Brokk.

'We will go to Asgard, you and I,' said Loki. 'I will bring Ivaldi's treasures and you can carry whatever three objects you and your brother can make. The three gods we will give them to, Odin, Thor, and Frey, will judge.'

'And if they choose me, I can have your head for my trophy?'

Loki laughed. Who could believe anyone could make greater wonders than those he held in his hands? 'Of course,' he said.

Brokk called for his brother Eitri and explained the wager. Eitri's eyes gleamed, for he too saw the chance of glory in the worlds above. He told Brokk to pump the bellows for the furnace. When the fire was glowing red he threw a pigskin into the middle of it.

'Keep pumping the bellows while I fetch what I need. You must not stop, even for a moment,' he told Brokk.

When Eitri had left Loki smiled to himself and stepped back into the shadows away from the fire's glow until Brokk could not see where he stood.

After a moment a great horse-fly came buzzing out of the shadows. It circled round, darted in, and bit Brokk viciously on the hand. Brokk cursed, spat at it, bent and licked the burning spot with his tongue, but kept on pumping. As Eitri's footsteps could be heard through the passages the horse-fly flew off into the shadows again.

Loki stepped forward to see Eitri and Brokk lift from the fire a boar, rounded, snorting, golden. Next Eitri placed a small amount of gold in the fire. When he told Brokk to pump without stopping again Brokk grumbled about the horse-fly and showed him his bite.

'Pump, brother,' said Eitri.

Again he went off down the maze of dark passages, again Loki blended with the shadows, again the horse-fly appeared. This time it went for Brokk's neck and bit him twice, once under each ear. Brokk hunched his shoulders and shook his head, but he kept on pumping. When Eitri returned he took from the furnace a golden arm ring.

For the third time Eitri threw something into the flames. It was a lump of iron. All happened as before. Eitri left, Loki seemed to disappear, the horse-fly appeared. It flew round as if choosing its spot with care. Suddenly it darted in and bit Brokk on his eyelids so that the blood ran down into his eyes, blinding him. Brokk lifted his hand from the bellows for a moment. He slapped at the fly and wiped the blood from his eye. Just then Eitri returned. He cursed his brother because he had stopped pumping. He reached anxiously into the furnace with his tongs and drew out an iron hammer, with a rather short handle.

'Look what you have done, fool!' he shouted.

'It's all right,' muttered Brokk. 'No harm has been done.'

'It isn't perfect,' said Eitri.

Still grumbling he wrapped the three things he had made and gave them to Brokk with whispered instructions. His eyes flickered round the cave, looking out for Loki. Loki did not reappear until Eitri had finished, and then he and Brokk set out for Asgard together.

Through Middle Earth they went: Loki, handsome and graceful, striding out; Brokk, short, dark, gnarled like a tree root that has broken out of the ground. When they came to Bifrost Loki carried Brokk so that

the red fire of the rainbow would not burn him. So they came to Asgard.

When all the gods and goddesses were gathered into the Hall Loki approached Sif. He placed the gold hair on her head. Immediately it was as if it had grown from her head. Sif patted it and smiled. Thor slapped Loki on the back and said,

'You have repaired the damage you did. But what presents have you brought us?'

Loki smiled and replied, 'I have brought you presents, and entertainment. You, Thor, and Father Odin, and Frey must judge whether my gifts are better than this dwarf's here. He and I have a bet to settle.'

Loki then presented Odin with the spear Piercer and explained that it could be thrown and would never miss its target. Odin held the spear thoughtfully. The two ravens on his shoulders fell silent. He sent them out at daybreak to fly about the world so that they could return and tell him all they saw. Now they sensed his mood and preened their feathers.

Odin sought wisdom, knowledge, and understanding. Perhaps it was because he was father of the gods. Perhaps it was because he had some warning of the last great battle between gods and giants and he wanted to learn more. He had already lost one of his eyes in this search. It was the price he had paid to drink once from the spring of learning that bubbles up from the root of the world tree, the Ash, Yggdrasill. Now he dreamed of how he could drag up from the roots of the world itself the deepest knowledge. He would hang nine days and nights on the Ash, his side pierced by this spear. By hanging from the tree, a willing offering, in pain and thirst, he could learn what he craved to know. Odin had followers in Middle Earth. Their quest for glory in battle and for wisdom often led them into madness. Odin had followers in Valhalla too, chosen dead warriors. These he kept in training for the last great battle. And so he brooded.

Loki was too pleased with his treasures to notice Odin's mood. He gave the boat to Frey. The gods exclaimed over Loki's wonders and then turned to Brokk. He stumped forward and gave the first object to Frey.

'This boar can run through the air or over the sea faster than any horse. Everywhere it goes will be as bright as day for its bristles glow as brightly as the sun.'

He then gave Odin the great gold ring, Dropper.

'Every ninth night eight other rings will drop from this one, all of gold.'

Odin started, for his mind was running on nines. The perfection of the circle and the purity of the gold chimed with his thoughts too.

Then Brokk produced the hammer and presented it to Thor. 'With this hammer you can hit as hard as you like and it will never break. If you throw it, it will always come back to your hand. If you wish to carry it, it will become small enough to fit inside your shirt.'

The three gods declared that Brokk had produced the best gifts, for with this hammer Thor could surely defend them against their enemies the frost giants.

Loki paled as Brokk smiled.

'I will give you gold for my head,' Loki said.

'No!' replied Brokk. 'I have gold enough. I want to take your head back under the ground as proof that my brother is the greatest craftsman.'

'Catch me, then!' cried Loki and ran off. Thor caught him before he had gone far.

'I will not have the gods dishonoured before anyone,' he said.

'Very well,' said Loki, his quick mind at work. 'Here is my head. Take it, but take none of my neck for that was not included in our bet.'

Brokk cried with rage for he saw that Loki had tricked him. 'If I cannot have your head, I will have your voice from within it,' he said. While Thor held Loki, Brokk used his awl to make holes in Loki's two lips and then lashed them together with a leather thong.

But Loki pulled the thong from his lips as soon as Thor released him and stood smiling crookedly with the blood dripping down his chin. Brokk went off, the empty-handed winner, feeling that he had lost.

Thor gently swung his hammer. 'I name you "Crusher" for you will crush the skulls of all our enemies,' he said. Thor used Crusher to guard Asgard but he also used it to bless his followers at birth, at marriage, and at their deaths. But Thor did not know that next time Loki caused trouble he would have to solve it without Crusher to help him.

Geirrod

LOKI thought it would be wise to leave Asgard for a while. He changed himself into a falcon and flew out over the worlds. It was in the land of the giants that trouble found Loki.

He had been flying over the bare rocks that border the land of Utgard when a blizzard caught him. Loki fought against it but his falcon's strength was not enough and he was carried, blindfolded by snowflakes, wherever the wind wanted to take him. As the storm died away he saw a tall tower standing black against the surrounding whiteness. He flapped towards it and perched on a high window ledge, exhausted.

Dark was creeping over the land and snow was still falling gently but a warm yellow light streamed out from the window where Loki rested. He moved further in, away from the dark and cold, towards the warmth and light. He looked down through the narrow gap.

Down below was a hall, full of feasting giants. Loki had become used to the elegance of Asgard and so the roughness of the scene below fascinated him. Half of the giants seemed drunk already, the rest were drinking hard. Food went into mouths, or on the floor if the eater was drunk enough. Some had rolled off their stools and lay snoring open-mouthed on the floor, a floor covered in scraps, spilt wine, and vomit. Huge dogs wandered about feasting on this litter, becoming drunk too.

As Loki stared down, the Lord of this hall, Geirrod, looked up and saw the falcon. He shouted to one of his more sober servants to climb up and catch the bird. It was a long and difficult climb up the rough walls and Loki leaned forward to watch. He meant to fly off at the last moment so that he could enjoy the servant's anger at having climbed so far fruitlessly. Perhaps the servant expected this and pretended to be stuck for a moment, perhaps Loki's wings were just too tired. When he tried to soar away he found his leg held and a cloth coming down over his head.

When the cloth was removed Loki found himself in a cage surrounded by the wide eyes and drooling mouths of drunk giants. Scraps of food were pushed in, and a bowl of wine, but Loki ignored them. Geirrod, who seemed less drunk than the others, said:

'This is no falcon. Look at its eyes! This is a shape-changer. Tell us who you are!'

Loki refused to answer. The cage was too small for him to change back into his own size and the prowling dogs made changing into something smaller unwise. Geirrod became angry at Loki's silence. He removed the food and wine and placed the cage on the floor.

'You will stay there,' he shouted, 'until you answer me.'

So Loki remained obstinately silent for three long months. Every night, as they became drunk, the giants would prod him. Every day the dogs tried to reach him through the bars with their long tongues. Always Loki remained silent, alone with his thoughts.

At last he was too weak to resist any more and he told the giant who he was. Geirrod held the falcon in his massive hands and said,

'I will release you into the air outside if you will promise to lead Thor here without his belt of strength or his hammer. If you will not promise you can return to the cage.'

Loki opened his mouth and promised with many binding oaths and so he flew away, free but with the burden of his promises. Loki returned to Asgard, taking as much time as he could on the journey. When he got back he was relieved to find that Thor was not there.

Thor was away from Asgard and none of the gods knew where he was, fortunately. He was visiting Grid, a giantess with whom he had become friendly. Although there was deadly hostility between gods and giants, from time to time friendships sprang up and flourished briefly before being frozen out by the disapproval of both sides.

Grid knew what Geirrod planned and warned Thor.

'Do not agree to anything Loki suggests for he will bring you into great danger.'

Thor laughed, and said, 'Loki is my friend. Some of his jokes do get out of hand, but he would never try to do me harm.'

'Take with you, then, my own belt of strength,' said Grid, 'and my staff and my iron gloves. Promise me you will go nowhere with Loki without them.'

Thor laughed again but he promised. He was not, therefore, very surprised when an embarrassed-looking Loki suggested a journey.

'I would leave Crusher and that great heavy belt of yours behind,' he said. 'They will only burden you where we are going.'

Thor agreed, easily, and they set off. Loki talked with nervous cheerfulness. Thor was thoughtful. He found it hard to believe that this was actually happening. His bluff mind could not reconcile Loki's friendliness and the doom to which he thought he was leading Thor. Thor thought he had been wise in not telling Loki of his conversation with Grid, nor of her belt round his waist, her gloves on his hands, or her staff he carried.

They travelled together until they reached the river Vimir. This flowed swiftly through rocky gorges, and was cold and deep. Thor grasped Grid's staff firmly and told Loki to hold his belt. He watched carefully where he placed his feet so that he would not lose his balance and get washed away. By midstream the water was up to their shoulders and rising steadily so that soon it would fill their mouths, and stop their breath. Thor looked up and saw, at the end of the ravine, Geirrod's massive daughter damming the narrow exit. He bent down through the

icy water and felt for a stone. He hurled it and it struck her full in the forehead so that she fell back and was drowned in the rush of the water. Thor and Loki were knocked off their feet too but Thor reached out and grabbed an overhanging rowan branch and so pulled them safely ashore.

As Loki led them to Geirrod's hall Geirrod himself came out to welcome them.

'Please wait in here for a moment,' he said, 'while I get everything ready for you.' He showed them into a dim room. It was empty except for a large wooden chair. Thor sat on the chair while Loki leant on the wall.

Suddenly Thor felt the chair and himself rising towards the roof, a solid wooden roof. He felt himself beginning to be crushed between roof and chair; he wedged Grid's staff between them and then pushed with all his strength. The staff straightened and the chair dropped. Two terrible screams were heard. On the floor were Geirrod's two other daughters who had been pushing up his chair, writhing with broken backs. Thor put them to rest with two blows of Grid's staff and pushed the bodies back under the chair.

Geirrod returned to find Thor sitting relaxed. Swallowing his surprise he invited them into his hall and suggested some games of skill before dinner. He threw a ball at Thor, a hard, difficult catch. As Thor returned it Geirrod snatched a red-hot lump of metal out of the fire with a pair of tongs and hurled it straight at Thor. Thor's hand seized it in Grid's iron glove. Geirrod hid behind a pillar in fear. Thor threw the iron with such force that it went through the pillar, through Geirrod, and through the wall of the hall. Geirrod collapsed on to the floor with the neat hole burnt through his chest.

Thor said nothing to Loki but walked off by himself, back to Grid to return her belongings. He told her everything that had happened.

'Do not think too badly of Loki,' she said. 'He was led into danger by enchantments and was forced to swear binding oaths.'

'I can never trust him now,' said Thor. 'He may have been trapped, but his own foolishness was at fault, and he said nothing to me to warn me. And even if he is blameless, trouble goes with him.'

While Thor stayed with Grid, Loki returned to Asgard, and trouble did indeed go with him.

The Apples of Iduna

Loki found Honir, one of Odin's young sons.

'Come,' he said, 'on a journey with me through Middle Earth and see what we can find to amuse us.' Loki was anxious to be away from Asgard when Thor returned so that his anger could cool.

Odin went with them. In his restless search for wisdom he found Loki's restless mind fitted his mood. Loki hurried them away without supplies of food.

'We shall find plenty on our way,' he said. 'Men are always honoured to feed us with their best.'

This was all right while they travelled through the populated lands of Middle Earth where they could be sure of finding a farm to stay at for the night. Later, as their journey seemed to lead them over mountains and deserts, they began to get very hungry. It was now inevitable that Loki should be blamed.

'This is typical of you,' said Honir. 'Heimdall the Watchman warned me as we left over Bifrost to be careful travelling with you.'

'Trouble goes with you,' said Odin, 'like a shadow.'

'You used to value my company,' retorted Loki. He was hurt and resentful. He felt he was unjustly blamed, and he was the more angry because he felt there was some truth in their accusations. He had, after all, hurried them away so that they would not be there when Thor returned.

They went on in an angry silence until, at last, they came down off the mountains into a green valley. A stream flowed through it and great oaks grew in the meadows. A herd of cows grazed peacefully and they had no difficulty in catching one. As they skinned it, jointed it and put it to roast over a fire their spirits rose.

'See,' said Loki, as they rested under one of the oaks and watched the meat roasting, 'all is well that ends well.'

When they thought that the meat must be ready they took it off the fire eagerly. To their surprise and annoyance they found it was hardly cooked at all. They put it back on the fire and waited impatiently. Loki felt that he was being blamed in some way for this too, and he sat apart, sullen and silent.

When they could wait no longer they pulled the meat off the fire again. It was still half-raw. This time their anger broke out into the open and ugly words flew about. They were interrupted by a voice from the tree above them.

'Your meat will never cook like this.'

They looked up and saw an enormous eagle perched near the top of the tree.

'If you give me enough to fill my stomach,' the eagle continued, 'then your share will cook, and cook quickly.'

'Of course,' shouted Odin, 'have all you can eat.'

The eagle swooped down and seized all the good tender joints.

This was too much for Loki's temper. He grabbed a branch from the ground and drove it hard into the eagle's side. It stuck there, quivering. The eagle turned its head, looked coldly at Loki, and flapped its huge wings. As the bird flew off, Loki found he could not let go of the branch. The eagle flew away northwards, low to the ground with one end of the branch stuck in its side, the other grasped in Loki's hands. Loki was dragged along the ground, through streams, over rocks, through brambles. He was torn and bruised and dazed.

'Let me go, let me go,' he begged as he was crashed into yet another boulder and through another thorn bush. Blood flowed from his cut face into his mouth and he spat it out.

'No longer hungry?' asked the eagle. 'Or is that blood not cooked enough for you?'

'If they were your cows, I apologise,' gasped Loki as the eagle flew low over the stream.

'I will not stop,' it replied, 'unless you swear to bring Iduna and her apples out of Asgard into the forest.'

Loki laughed bitterly. 'I cannot take a step out of Asgard now without having some terrible oath demanded from me. It is no use trying to fight against fate. I swear to do as you wish,'

Immediately the stick fell from the eagle's side and from Loki's hands. The eagle flew away and Loki limped back to his companions. His jolted mind was fill with pity for himself and fear for what he had to do.

Not even leading an unarmed Thor into Geirrod's hall was as terrible as the task he now had. He had to cause the death of all Asgard. Iduna's apples kept the gods young. If they did not eat them they would grow old and feeble, and die.

Loki stumbled back to find Odin and Honir full and relaxed. They had left very little for Loki and they laughed at his appearance.

'I see you have been changing your shape again, Loki,' said Odin. 'Let me see if I can guess. Yes, I know, a dung-heap.'

'Didn't you pluck your chicken, then?' asked Honir. Loki did not reply but he felt a fierce joy growing in him. He would revenge himself on all those who had slighted and mistrusted him. He began to plan.

A few days later an excited-looking Loki burst in on Iduna.

'Show me your apples, please.'

Iduna, puzzled, opened the box that she kept them in. Loki was the child of giants and had no need of the apples. Loki pretended to examine the small green fruit intently. Then he said,

'I am almost sure I have found a tree that produces apples just like these. Think what that will mean! There will be no danger of Asgard ever running out. Bring the box quickly so that we can compare them.'

Loki talked excitedly and half-dragged Iduna out through the great gate of Asgard and into the surrounding forest. There was the beating of mighty wings, and Iduna and her apples were gone, gone to Utgard, the land of giants.

Loki looked thoughtfully at the rapidly disappearing eagle. How long would it be before the gods realised Iduna had gone? What would happen to them without the apples? He went slowly back to Asgard.

Because the gods had been kept young by the apples of Iduna they started to age very quickly. Thor had wrestled with Old Age in Utgard and found her too strong for him. The magic of the apples had kept her away so that the fight could not begin. The gods could die but until struck by enemies without or traitors within they remained young. Now their faces became lined. Their hair became grey and thin. Their backs became bent. Their movements slowed. As they realised what was happening they hobbled round Asgard searching for Iduna, questioning each other, worrying endlessly and ineffectively.

So things might have continued, until Old Age finally bore them to the ground and won, had not Heimdall's keen ears heard talking in Hel. Heimdall was the lookout at the rainbow bridge. He could hear grass growing in the fields of Middle Earth, and the wool growing on the sheep eating that grass, and the fleas moving in that wool. Now he heard Hel laughing, Hel the gloomy Queen of the land of the dead.

'Soon, soon now, all of Asgard will be down here in my Hall. All those proud gods who exiled me here will be at my command. Odin can count the bones and Thor can hammer them flat.'

Heimdall tottered into the Hall of Asgard and told Odin what he had heard. Odin roused himself and looked round at the white hair and toothlessness of those who had, just weeks ago, been young and handsome. One of his ravens leant down and cawed into his ear, 'Loki'.

Odin spoke, and his voice was not much better than the raven's croak. 'Fetch Loki!'

Loki was dragged into the Hall, still his old self for he was not a god. He was overawed by the desolation and despair around him. He had always wanted the love and respect of the gods and he still hoped they would accept him, even now. He fell to his knees before Odin.

'Father Odin, forgive me. A promise was forced from me. I did not know what would happen to you without Iduna's apples.'

'Forgive you?' said Odin. 'You want the forgiveness of a dying god? What good is forgiveness? Undo what you have done or I will utter my curse upon you. Return Iduna to us and I will never curse you.' Odin did not know how soon he would regret his promise.

Loki turned himself into a falcon and flew off towards Utgard. He had learnt that Iduna had been taken by a giant called Thiazi who had the power to turn himself into an eagle. He flew to Thiazi's hall and perched wearily on a window ledge. With his falcon's eye he saw Iduna sitting sadly in the hall while Thiazi was rowing on the sea with his daughter. Loki swooped through the window and perched on Iduna's shoulder. Now he could use his power of changing to turn her into a hazelnut. He grasped the nut in one talon and set off back to Asgard.

When Thiazi returned from his fishing he saw at once that Iduna was gone, and saw that all his doors were still securely barred. He took on the shape of a giant eagle again and flew swiftly and strongly after Loki.

Heimdall the Watchman, with his eyes that could see the crabs crawling on the bottom of the sea, saw first a falcon, tired, slow, with a nut clasped in its talons. Behind he saw an eagle, gaining with every wing-beat. Hurriedly he ordered the gods to build great piles of wood-shavings along the walls of Asgard. So it was that, as Loki dropped exhausted straight down, the gods set fire to the bonfires. Too late the eagle saw the flames leaping up. His speed carried him into the heart of the fire and he fell blazing to earth and was burnt.

'I have cooked you at last,' shouted Loki. He turned Iduna back to
her own form. She gave apples to the gods. They bit into them and as the
juice ran down their throats they grew young again. They went cheerfully
back into the Hall of Asgard.

Thiazi's daughter, Skadi, learnt what had happened to her father.
She put on her armour, seized her weapons, and went to Asgard for
revenge. When the gods received her kindly and spoke softly she was

puzzled. She looked at the warmth and happiness around her and envied it, for she had never once laughed in her life.

'What can you offer me in place of a father?' she asked.

Skadi was beautiful but gloomy. Odin looked at her and said,

'We will give you a god to be your husband.'

At this Skadi's heart lightened. Her eyes turned towards Baldur, the most handsome of the gods. Baldur's mother, Frigg, hastily said,

'All the gods will take off their shoes and stand behind that screen. You can choose your husband by his feet.'

Skadi looked along the line of feet that stuck out under the screen. She stopped in front of the most perfect pair, thinking they must be Baldur's.

'I will have him,' she said.

When the screen was taken away it was not Baldur who stood there but Njord. Loki saw Skadi's face and shouted out before trouble could break out for which he would be blamed:

'Lucky, Skadi! A handsome young husband, with a most loving heart!'

Skadi had suffered long from her father's coldness and unkindness. She turned and looked more kindly on Njord.

Loki felt triumph growing in him and said again, 'I will make you do what you have never done before.' He turned to Honir:

'Quick, fetch one of Thor's goats, fetch the lame one.'

'Fool,' said Thor, 'do you not remember that I have lost that one?'

'Thank goodness I was not riding it then,' said Loki, 'or I would be lost too. Fetch the other one then.'

When the goat was led in he held it by its horns and leapt on to its back. There he sat, facing the goat's tail. Skadi started to smile but Loki shouted at her.

'What are you smiling at? Is it not sad that mighty Thor has one lost lame goat and this beast?'

'What is wrong with it? It looks a fine animal to me.'

'It may look fine to you,' replied Loki, 'and it would be fine for one-handed Tiu but surely you can see that if when I mount it I have to sit backwards it must be a left-handed goat.'

Skadi laughed, who had never laughed before. Her heart was light. She had a handsome young god to replace her surly father. The red-haired Loki was riding round the room backwards on a goat just to amuse her. To please Skadi further, Odin then took her dead father's eyes and threw them into the sky to make stars. Skadi departed, happy.

'Do you not see, Loki,' said Odin, 'how kindness and wisdom are the best replies to trouble, not deceit and spite?'

Loki laughed, but looked sideways at Baldur. It was a pity Skadi had picked the wrong feet.

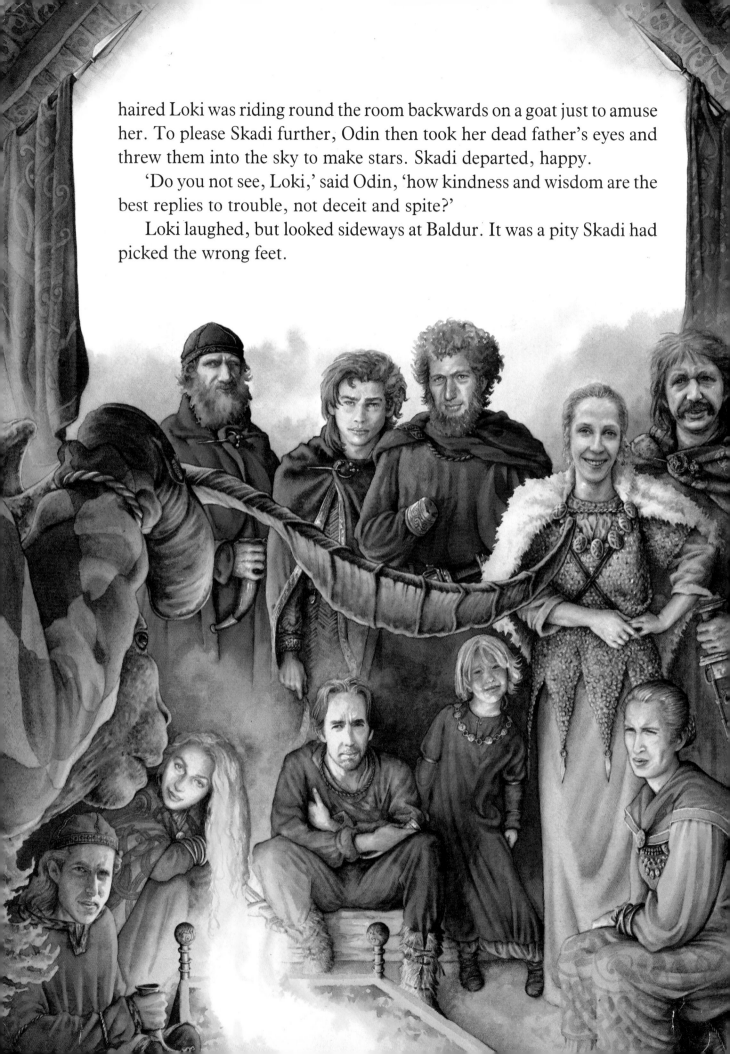

Baldur

BALDUR dreamed. Dreams invaded his sleep so that he woke trembling and cold. The gods were not used to nightmares and Baldur should have been the last to suffer them. Baldur was the wisest and best-looking of the gods. He was so gentle and kind that nothing evil could live near him. He seemed to shine out, even among the gods.

Loki was, of course, jealous of Baldur. Loki was the second handsomest in Asgard. Loki was clever rather than wise. Loki was not gentle or kind. Evil seemed to grow up near him. His jokes and tricks went wrong and even his friends had grown cold towards him.

If Loki had had bad dreams no one would have been surprised, least of all Loki himself. Perhaps he did, and kept quiet about them. Baldur was surprised and tried to describe his dreams to the other gods. All he succeeded in doing was to worry them with forebodings of disaster. He painted images of flickering red, like fire, like serpents. He described feelings of piercing, like roots growing into his body. Above all he filled their minds with the suffocating blackness that smothered him in his sleep. Fear swallowed them up. No one knew what they were afraid of, but they were all afraid for Baldur. The most afraid was Frigg, Odin's wife, for she was Baldur's mother. She questioned him and questioned him about his dreams but all he could talk about was flickering red, piercing, blackness.

When Frigg talked to Odin, the All-Father was vague and said nothing to calm her. He talked of Fate and of the Well of Urd.

The great Ash Tree, Yggdrasill, had its roots in three springs, one each for the worlds of the gods, the giants, and the dead, Asgard, Utgard and Hel. By the spring in Asgard, which was called the Well of Urd, sat the three maidens who wove the fate of the worlds, the Norns.

Frigg could not sit and do nothing. She left Asgard and went through the worlds. She made everything she met swear an oath that it would not harm Baldur. Everything she asked swore that it would do Baldur no harm, and swore readily because he was loved by all—all except Loki.

Frigg went first to anything that might have been in Baldur's dreams: the red flickering flames of fire, the shimmering coils of serpents. She went to everything that could pierce: to the trees whose wood makes spears, to the iron which the dwarfs beat into swords, to stone which anyone can make jagged and throw. She then went to likely and unlikely things: to grass on which he might slip, to water which could drown him, to fog in which he could lose his way and wander until exhausted. All swore that they would not harm their golden god.

Frigg returned to Asgard. At dinner that night she told Baldur that he had nothing to fear.

'Everything in all the worlds has sworn not to hurt you. Dream no more, my son.'

'Let's try then,' said Loki and threw his drinking-horn straight at Baldur's head. The horn fell to the ground before it could touch him. Loki, smiling outwardly though raging inside, grabbed everything within reach and hurled it with all his strength straight at Baldur. It all fell to the ground without touching him: plates, jugs, knives, stools, all fell harmlessly. A feeling of excitement and release from tension swept through the gods and they all hurled things at Baldur while he sat smiling happily. This became an evening's entertainment for the gods when they had drunk enough, but nothing ever touched Baldur enough to hurt him. Loki was hurt, though, as Baldur became the gods' chief entertainment.

Baldur did not tell his mother that his dreams continued for he believed that she had done everything that she could. Frigg did not ask for she was reassured every evening by the game of Aunt Sally that the gods played.

But Loki noticed Baldur's haggard face in the morning—and noticed how he smiled when his mother was near.

One evening after dinner in the Hall of Asgard an old woman hobbled in and sat down next to Frigg. She started talking of her children and of all she had done for them. Then she watched the gods throwing at Baldur and the conversation seemed to turn inevitably to Frigg's journeys to make him safe.

'And did you really visit every single animal, and plant, and . . . and . . . thing?' the old woman asked. 'My goodness, that was a task indeed. How long did it take you?'

Frigg described her journeys through the worlds and talked of the oaths she had received from everything.

'And did you ask every last thing in all the worlds?' probed the old woman.

'Well, no,' said Frigg, lowering her voice and glancing round to see if Loki was near. 'There was a small bush growing west of Valhalla called mistletoe that was . . . well, I thought it was too small and young, and just too far away to be any threat to Baldur.' Frigg did not say that she feared the ancient magic of the mistletoe that grew green and flourishing amid the winter deadness of the great oak tree.

'Did you really visit Valhalla?' interrupted the old woman, rather hastily, as if she wanted to change the subject. 'Surely no woman is allowed there? What is it like?'

'You must remember that Odin, my husband, is Lord of Valhalla. That is why I was able to go there on my quest. It has a wonderful great Hall roofed with shields and there go all those who die in battle. The Hall has more than six hundred and forty doors for I lost count there. Each door is wide enough for nine hundred and sixty men to march out of to battle.'

'Who can they fight in that dead land?' asked the old woman.

'They fight each other all day until they are all killed. In the evening they all come to life again and go back to the Hall for feasting, drinking, and long stories of their greatest battles. Odin told me,' Frigg said, leaning forward and speaking confidentially, 'that he is keeping these warriors in training for the last great battle.'

'But what can they find to feast on in that dead land?'

Frigg did not like the way the old woman kept talking of the dead land but she answered, 'They eat roasted boar which, like Thor's goats, come back to life again each morning. Their drink is the most intoxicating mead. It flows from the udder of a goat that stands all day on its hind legs and tears at the buds on the great tree that grows there. So much mead flows from her that there is more than enough for all the warriors.'

'It sounds better than the dread land of Hel,' said the old woman, 'and that is where those who do not die in battle will go.'

As Frigg had spoken of Valhalla she had forgotten her indiscretion in the pleasure of remembering the sights she had seen. The old woman said farewell and hobbled out of the Hall again and Frigg smiled fondly at the gods amusing themselves throwing at Baldur.

Loki was not seen in the Hall of Asgard for several days. The first god to know of his return did not see him at all, for it was blind Hod who heard him first. Blind Hod stood smiling while around him the other gods threw at Baldur.

'Why don't you throw something?' whispered Loki in Hod's ear.

'I have nothing to throw,' Hod replied. 'And, besides, how would I know where to throw?'

'I will help you,' whispered Loki. 'here, hold this twig.'

Loki placed his branch of mistletoe in Hod's eager hand and turned

him until the blind god could aim straight at Baldur.

'Now!' he whispered.

At that moment Baldur looked up and saw through the crowd the light flickering red on Loki's hair. A terrible suspicion flashed into his mind. But Hod had thrown the mistletoe, sharpened by Loki, and it went clean into Baldur's body. Baldur felt it pierce his chest like roots growing into an oak tree and then . . . blackness overwhelmed him and he dropped like a felled oak, dead.

Slowly, silently, the gods encircled Loki. One voice only was raised in grief: Frigg grieving for her son. The rest put aside grief for the moment as they remembered the years of Loki's troublemaking.

Loki looked from one to another, his quick mind searching for weakness to exploit. He turned first to Thor.

'You cannot blame me for this terrible accident. How could I have known what would happen? Everything else dropped harmlessly from our dear Baldur. Look at the heaps around him. I saw this plant west of Valhalla, its green bright against the brown of the oak it grew upon. I thought of Baldur, bright against the darkness of the world.'

'Go!' said Thor in a voice of thunder.

Loki looked round and saw from the faces of the gods that he must go, and go quickly without another word. He walked out and the gods let him go.

Then Frigg lifted up her voice and cried out, cursing herself because she had not made the mistletoe promise not to hurt Baldur. Tears streamed down her face and she shook with grief.

'Do not blame yourself,' said Odin, her husband. 'It is the Norns, Past, Present, and Future, who sit by the spring that wells up under the great Ash Tree Yggdrasill. They weave the fate of men and gods alike. There is no way you can avoid what they have decided.'

Frigg could not accept that there was nothing she could do.

'Who will ride to Hel and beg for the soul of Baldur to be returned to us? If we offer Hel fair words and a ransom surely she will release him?'

Her son Hermod stepped forward. He was brave enough to undertake that terrible journey.

'Ride my horse, Sleipnir,' said Odin. 'His speed will get you there with less danger, and we will have news more quickly.' Odin remembered that Hel was Loki's daughter and feared that she would not undo her father's work.

So Hermod leaped onto Sleipnir's back and rode swiftly out of Asgard. When he had gone the gods prepared for Baldur's funeral. Frigg washed away the blood and dressed him in his finest clothes. They carried his body down to the shore and laid it in his ship, Ringhorn. Thor blessed the ship with his great hammer and Odin placed on Baldur's arm the gold ring, Dropper, that Loki had brought from the dwarfs. They laid his sword by his side and prepared to push the ship out on the water so that it would burn at sea. However hard they tried they could not move the ship for it was too heavy. At last they sent for a giantess. She came riding on a great wolf, using adders for reins, and it took four of the gods to hold her beast. She grasped the back of Baldur's ship and gave it a heave that made it run down to the sea so fast that the rollers caught fire and the earth shook. A great crowd watched Baldur's funeral fire, with its flickering red flames and pall of black smoke. Their hearts were pierced with grief.

All this time Hermod was riding the eight-legged horse Sleipnir towards Hel, the land of the dead, and its fearful ruler. Even with Sleipnir's speed it was a journey of nine nights, northwards and downwards, ever downwards. Hermod travelled through deep gorges, dark ravines, and valleys where the sun never reached. At last he came to the Resounding River that flows swift and black between the worlds of the living and the dead. Hermod rode along its bank until he came to the Echoing Bridge that spans it. As he crossed, the shrivelled guardian of the bridge came out of her hut and stopped him.

'Who are you riding across the bridge? You make more noise than fifty of the dead. Why are you riding to Hel? You do not look ready for that kingdom yet.'

'I am seeking Baldur. Has he passed?'

'You are too late to catch him,' the old hag replied gleefully. 'He passed over my bridge with a great host of the dead. By now he will have passed through the Gates of Hel. They will not open for you.'

Hermod urged Sleipnir on. They galloped through the mist and darkness that fills all that land until the great walls of Hel loomed up blacker than the surrounding dark; walls sealed with forbidding gates. Sleipnir's eight legs carried him over the Gates of Hel and so Hermod came to the kingdom of the dead. It was cold, cold enough to numb his mind as well as his body. It was wet, wet enough for the ground to clutch at his feet. It was noisy with unseen noise: the howls of the hounds of Hel

and the cries of the dead. Hermod had to push through cold and dark and fear to the very doors of the Hall of Hel. He opened the doors and stood in that dark, damp, sad Hall. He saw, sitting on a raised seat, Baldur still, silent.

Then Hermod's mouth was filled with the stench of rotting flesh as Hel herself stood before him. She had grown, and grown fat, since Odin had thrown her as a baby out of Asgard.

'What do you want in my kingdom?' she asked. Hermod wilted at the blast of her poisonous breath.

'I come to beg for Baldur,' Hermod said simply, too cold and fearful to flatter as he had planned.

'Yes,' she replied, and her smile was the worst thing Hermod had yet seen, 'Baldur may return to Asgard if everything you meet on earth as you journey back weeps for him. I will know if you are successful, and I will send him.' She turned to Baldur, 'Give Hermod some token so that Odin will know Hermod speaks true.'

Baldur reached out his cold white hand and dropped the gold ring Dropper into Hermod's palm. Then his hand fell slowly on to his lap again and so he sat, still and silent.

Hermod stumbled from the Hall and dragged himself onto Sleipnir. This time the Gates of Hel swung open for him and he could ride out into what had been the terrors of the downward journey. As he came out into daylight he called out, 'Weep, weep! Weep Baldur out of Hel.' He saw with wonder that every plant, every tree, every stone wept. They were dripping as they do when ice melts in Spring. Hermod rode on with a light heart, sure of success. Every man, woman and child he passed wept as he cried out to them.

At last, near the borders of Asgard, he was riding through a dripping forest when he came to a cave, and in the cave was an old woman, dry-eyed.

'Weep for Baldur, weep him out of Hel,' Hermod called, thinking the old woman ignorant of what had happened.

'Why should I weep? Baldur was no use to me. My eyes will never water for him. Let Hel keep what she has.'

At these words Hermod's heart broke and he rode weeping through the weeping worlds to Asgard.

The Binding of Loki

WHEN Hermod had told Odin all that had happened and had given him Dropper Odin sat deep in thought. Then he roused himself and said:

'This is Loki's doing, all of it. Frigg told me of the old woman who came and talked to her and learnt the secret of the mistletoe. Now that same old woman will not weep Baldur out of Hel. That old woman is Loki, the Shape-Changer, and Hel is his daughter.'

'You are right, Odin,' came a voice from the doorway. There stood Loki, smiling his crooked smile defiantly round the Hall. 'You are right, but you cannot touch me here. This is a holy place. And you have sworn you will never curse me.'

'You should weep for Baldur,' said blind Hod. 'You guided my arm and gave me the mistletoe.'

'Oh dear, oh dear,' said Loki, 'are you blind in mind as well as in your eyes? Did you not wonder why only I helped you throw?'

Honir laid his hand on Hod's arm. 'Only the evil look for evil in other's actions.'

'That's good from you, Honir,' said Loki. 'I can remember when all you wanted was to fill your own belly. You didn't care that I was dragged off by a monstrous eagle. Your eyes are never off your plate.'

'Loki, have you no respect?' asked Sif. 'Poor Baldur is newly dead and all you do is abuse his family.'

'All you do is peer in your mirror. I'm surprised you've even noticed "poor Baldur" is dead. Oh well, it's one admirer less I suppose.'

'That is enough,' said Tiu as Sif wept.

'What do you owe this gang?' asked Loki. 'You were the only one brave enough to lose a hand to protect them and what reward have they ever given you? They have even stopped listening to all your stories of your heroic deeds.'

'Peace, Loki,' said Iduna. 'We do not measure reward and punishment like a miser counting copper coins and for that you should be thankful. You have never paid the price for all that you have done.'

'You may well talk about copper coins,' replied the angry Loki. 'You who carry round a box of withered apples like a peasant girl at the market.'

Heimdall stepped forward to rebuke Loki but Loki turned on him before he could speak. Heimdall's steadfastness as the Watchman of Asgard had always been a reproach to Loki's careless ways.

'I suppose you're going to stick your nose in now. Why don't you get back to your precious rainbow and look after that? It shows what Odin thinks of you that you have to sit forever with your back getting wet and your eyes staring into space and your tin trumpet rusting at your side.'

One by one Loki abused each of the gods. Words spewed out of his mouth: a mixture of lies and filth and, most hurtful of all, truth. The gods became more and more enraged and hurled insults back. Loki's quick tongue ran away with him and found wounds to irritate in all of them. At last Odin, sickened by the unseemly brawl that had developed, shouted:

'Loki, my wife told me of the old woman who came to talk to her. You planned all this in the malice and cunning of your heart.'

'Yes,' replied Loki. 'Once, Odin, you swore that we were blood-brothers. You swore that you would never drink unless we drank together. Now look at you, taking their part against me! But you have never been known for your honesty and fairness. You allow cowardly warriors to win in battle. Half the time you are lost in your daydreams, pretending to be wise. We will see how wise and brave you are when you face Fenrir the wolf in the last great battle.'

'Loki, what did I ever do to you that you have taken my son from me?' asked Frigg.

'If you had not been too frightened of the mystery of the mistletoe in Valhalla it could never have harmed your precious son,' replied Loki. 'Blame yourself.'

'Loki, you have said too much,' shouted Thor. 'The less you say now the better.'

'The less I say the better for you,' replied Loki. 'Your slow wits limp along behind my words like your lame goat. It wouldn't be so bad if you were just stupid but you are cowardly as well. I remember you shivering with terror in a giant's glove.'

Thor raised Crusher in great anger but Frigg laid her hand on his arm. She did not want the holy place defiled. Slowly Thor lowered his hammer. Loki knew that he had said too much and that he must go. He turned and left the Hall of Asgard for the last time.

Loki knew that the gods would seek vengence and he was afraid. He fled to a high mountain and there built himself a shelter with four openings, one facing in each direction so that he could keep a good look-out. Here he spent his time, partly in remorse for what he had done, for the place in Asgard he had thrown away, and partly in planning his escape. He knew he must be found in the end for the gods would search until he was. He turned himself into various shapes and tried them out, seeking safety. In the end he chose the shape of a salmon and practised swimming down the mountain streams. In the evenings he sat and

brooded while fear of the gods grew in his mind. To escape the hunter you must think like him. How could a salmon with the mind of Loki be caught?

As he sat and brooded his mind went over and over the past. He found himself weaving grasses as he had once sat by the fire and woven Sif's hair. His clever fingers wove a net, the first net ever made. He looked at it, saw it could catch a fish by the gills, and burnt it in horror. So his own cleverness caught him.

When the gods found his hiding place they searched even the ashes of the fire and saw the pattern of the net. Honir realised that such a shape could catch fish. He quickly copied the design and ran with Thor down to the river. There was a fine salmon. They hid the net they had made and walked downstream, pretending to be looking on the land for Loki.

They came to a waterfall near the sea and stretched the net across the river below it. Honir then went back and drove the fish down the river. Loki leapt down the waterfall and, too late, saw the net straight in front of him. The force of the water and his own speed made it impossible for him to stop. With a mighty flick of his tail he leapt over the net, but Thor was too quick for him. Loki found himself grabbed just above the tail. Suffocating in the air he turned himself back into his real shape.

The gods took Loki back to the cave he had sat in dry-eyed. They took three slabs of stone and drilled a hole in each. They then set them upright in the earth. They fetched Loki's two sons and turned one of them into a wolf. Immediately the wolf turned on his brother and tore him apart. They chased the wolf away and gathered up the dead boy's intestines. With these they tied Loki to the stones: at shoulders, waist, and knees. As soon as his son's intestines were knotted around him they turned to iron. The gods then found a venomous snake and hung it up above Loki's head so that the poison dripped steadily, drop by drop. Loki's wife sat at his head holding a basin to catch the poison. Every time the basin became full she had to empty it and then the drops splashed into Loki's eyes and ran down his face like tears, the tears he would not weep for Baldur. Then he would shudder and be convulsed so that the whole world was shaken by an earthquake.

Loki lay bound under the dripping snake, tied by the wolf-torn entrails of his son, in his own Hel. Here all love for the gods was burnt out

of him as the poison of hatred filled his mind and revenge was his only craving.

As Loki lay in his cave the worlds changed. Perhaps the death of Baldur and the gods' vengeance had tipped the balance, the balance between summer and winter, good and evil, gods and giants. Perhaps the earthquakes that shuddered through the worlds from Loki's cave shook loose the bonds that held things together.

And Loki lay waiting, waiting for the last great battle, for Ragnarok.

Ragnarok

WARS spread over the earth. Greed and fear and pride drove all men to terrible slaughter. All human kindness and warmth was frozen out. The worlds were falling apart; everything had gone wrong.

There were three winters of war and three winters that ran into each other without a break. It was a time of driving snow carried on bitter wind, of darkness and hard frost. In the third winter two mighty wolves ran through the worlds, wolves fathered by the chained Fenrir. They lifted up their grey muzzles and howled at the light. One snapped at the moon and swallowed it up. The second wolf then stretched out his neck, opened his mouth wide, and slobbered down the sun. Darkness swept over the lands and winds shook the earth so that forests fell and mountains trembled.

Then, as though the cold darkness were not enough, the three terrible children of Loki broke upon the worlds. Fenrir the Wolf was freed from his bonds by the earthquakes. As he swam across the grey sea towards the mainland he howled out to his brother Jormungand, the great serpent encircling Middle Earth. Jormungand heard him and writhed about, coiling himself up ready to slither on to the land. His movements produced tidal waves that crashed mountains of salt water onto the shores and down to the borders of Hel. Here Loki's daughter launched the Ship of Death, the ship made out of dead men's nails. She boarded it with Garm, the hound of Hel and a crew of frost giants. So Loki's three children prepared to avenge their father's shame.

Fenrir's great teeth tore off Loki's iron bonds and released him. Loki rode on Fenrir's back down to the great water and found there Hel waiting for him.

Fenrir ran foaming at the mouth towards Asgard and his old enemies, his lower jaw touching the earth, the upper the clouds. Jormungand whipped the waves and spewed out his poison so that it rained down from the sky. Fire flashed out from their nostrils and lit up the darkness. The Ship of Death steered by Loki sailed on the floods bearing Hel and her terrible crew towards Asgard.

Between Middle Earth and Asgard stretched the rainbow bridge, Bifrost. Here stood Heimdall, the Watchman. He strained his eyes through the blackness around him. Strange howlings and cries filled the air, the noise of great waters, the screech of the gales. Gradually the darkness was lit up with flashes of red: red fire from the mouths of wolf and serpent. Heimdall put his horn to his lips and blew a great blast to summon all the gods, the trumpet blast that announced the end of the worlds.

Heimdall had been waiting for this moment since the great darkness came. As the time of wars and winter had come to the worlds Odin warned that the Last Days had come; they must prepare for the great battle.

Odin now summoned the gods to him and called up his host from Valhalla. The gods armed themselves in the flickering firelight and set off into the darkness with the ghostly host beside them. The giants, too, collected in Utgard under their Lord. So they came down into the plains and the night was broken by the torches of the approaching armies of gods and giants.

As the Ship of Death sailed on the flood waves towards Asgard the last great battle was joined on the plain. The armies swept to and fro with Odin and the Lord of Utgard leading them. Many giants were slain and it began to look to Odin as though the gods could win when, with a howl that froze the gods' minds, Fenrir bounded forwards. Odin made for Fenrir with Piercer, the spear the dwarf had made him. Fenrir opened his mouth wide and Odin rode the red path of the wolf's mouth and was

slashed to pieces by the white swords of his teeth and swallowed. But before Fenrir could get the Father of the Gods to his stomach, Hermod ripped his throat apart and god and wolf were strewn over the plain.

Thor raised Crusher over Jormungand's head and smashed it down. It went straight through and embedded itself in the earth beneath. The great serpent's poison-sacs burst and Thor went down spluttering and choking in the burning liquid that flowed over him. And so Thor was burnt as on a funeral pyre by the flames from the dying Jormungand.

Tiu had fought like ten with his one hand. While Odin and Thor were grappling with wolf and serpent he was attacked by Garm. The hound of Hel fought with him bitterly. They were so evenly matched that they

killed each other. And so the battle raged over the plain in the icy
darkness so that few could be sure whom they fought.

Heimdall, the steadfast watchman of Asgard, was unrelenting in his
search for Loki. He had seen all the trouble Loki had caused and was
determined he should not escape. At last he came face to face with him.
There was a pause as the two bitter enemies gathered their strength. Then
their swords pierced each other's heart and they fell together in death.

As Loki fell so from out of the south rode the sons of Muspell from the land of fire. The darkness fled and flame, too bright to look on, filled the sky like a thousand suns. As the fire and lava flowed white-hot over the worlds the mountains melted into the sea. Asgard, Utgard, Middle Earth, all was carried by the molten rock under the surface of the sea. As the fire swept away, darkness returned and sat on the tossing water. All else had gone. That was the end, but it also prepared a new beginning.

IN time a new sun arose from the fiery seed sown by the sons of Muspell. The new sun gave birth to a new moon and so light returned and made day and night. In time a new land rose out of the waters and grew green and fresh. In time birds and beasts covered the land with life and movement. In time new men and women spread over the land and a new Father ruled the heavens.

And Loki? His heart and spirit were shattered into small pieces at his death. Many of us today have a little of him in us, to our cost.